Tiling

by M...

Table of Contents

Tiling Shapes 2
Squares 4
Triangles 6
Hexagons 8
Trapezoids 10
Fun with Tiling 12
Glossary 15
Index 16

Consultant:
Adria F. Klein, Ph.D.
California State University, San Bernardino

capstone
classroom

Heinemann Raintree • Red Brick Learning
division of Capstone

Tiling Shapes

Look at the **squares** on this game board. They fit together perfectly. There are no spaces between them at all. This is an example of **tiling**.

These building blocks are tiling shapes. They fit together just right, with no spaces between them. Tiling shapes are all around us. Read on to find out more about them.

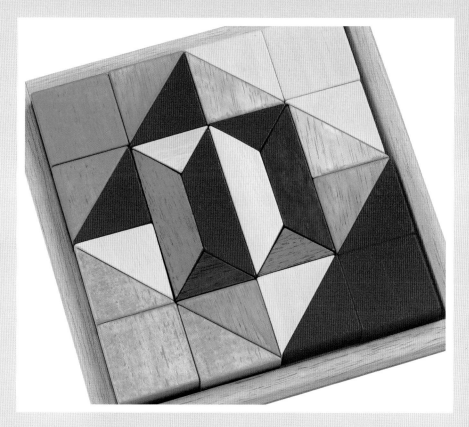

Squares

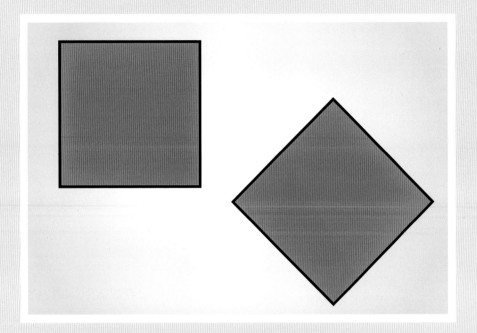

Think about a square. It has four sides that are all the same. You can **rotate** a square, and it will still look the same. A square is an example of a tiling shape.

This quilt is made of tiling squares. There are smaller squares inside of larger squares. All of the squares fit together perfectly, with no spaces in between.

Triangles

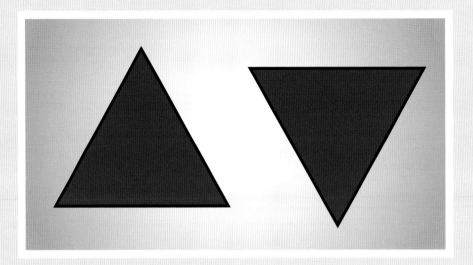

Triangles like this one have three sides that are just the same. You can rotate or **flip** this triangle. But because they are **equal**, one side will always match up perfectly with another.

These triangles fit together just right. You can put them together with no spaces in between. You have to turn the triangles in different directions to do it, but you can make them fit.

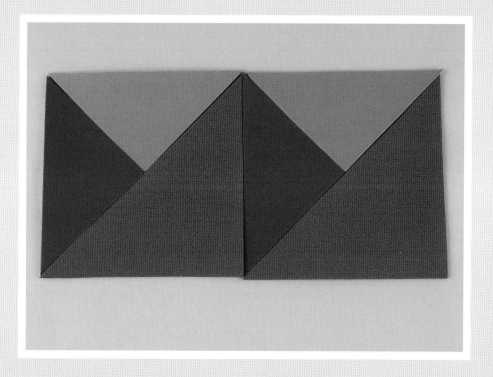

Hexagons

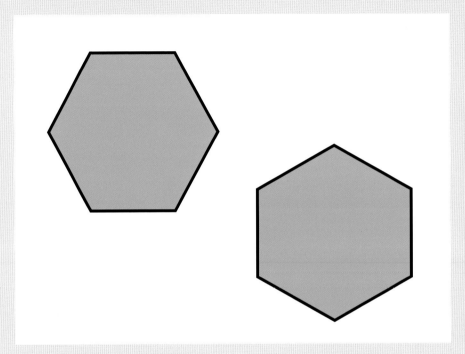

A **hexagon** is a shape with six equal sides. You can flip or rotate a hexagon, but it will still fit together perfectly with other hexagons.

This is the inside of a beehive. It is an example of tiling shapes in nature. These hexagons fit together perfectly with no spaces between them. It seems bees are good at tiling!

Trapezoids

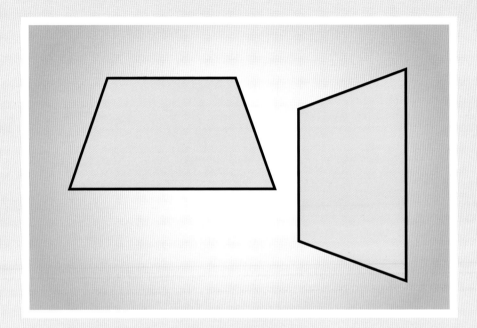

A **trapezoid** is a four-sided tiling shape. Its shape stays the same when you rotate or flip it. You can fit two trapezoids together perfectly, with no space in between.

If you put the longest sides of two trapezoids together, you make a hexagon. Hexagons and trapezoids make good tiling shapes.

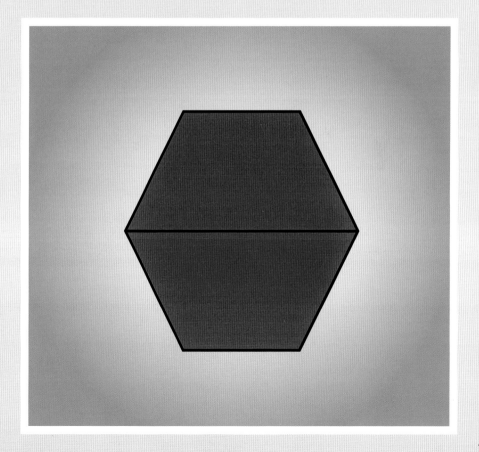

Fun with Tiling

Look at the shapes in this photo. Squares, triangles, and rectangles are tiled in this building. Try to find examples of each.

Here is another quilt. It has many tiling shapes in it. See if you can name them. Then look for smaller shapes that have been put together to make larger shapes.

Tiling shapes are everywhere. Once you know what to look for, it's easy to find them. They're on clothing, in games, and maybe even in your bedroom!

Glossary

equal the same as
flip to turn over
rotate to turn at a fixed point
tiling covering with tiles

Index

blocks, 3

flip, 6, 8, 10

hexagons, 8, 9, 11

quilt, 5, 13

rectangles, 12

rotate, 4, 6, 8, 10

squares, 2, 4, 5, 12

trapezoids, 10, 11

triangles, 6, 7